TO BEERSHEBA 1917

WITH THE AUSTRALIAN LIGHT HORSE

COMPILED BY TOM THOMPSON

With Photographs
from the Haydon Family Archive
Text by Guy Haydon & Ion Idriess

ETT IMPRINT

Exile Bay

Celebrating the Australian Light Horse Charge at Beersheba

This Imprint Classics edition published by ETT Imprint, Exile Bay 2024

First published by ETT Imprint in October 2017
Colour edition 2020

Extract from *The Desert Column* by Ion Idriess (ETT Imprint 2017)
Letter from Guy Haydon courtesy Peter Haydon

ETT IMPRINT
PO Box R1906
Royal Exchange NSW 1225 Australia

ISBN 978-1-923205-02-4 (paper)
ISBN 978-1-922384-92-8 (ebook)

Design by Tom Thompson

A Note on the Haydon Family Photographs

When war broke out in 1914, Guy Haydon enlisted in the 12th Light Horse with his younger brother, Barney, and their horses, Midnight and Polo. The two brothers were sent to Gallipoli at the end of August 1915, leaving their horses in Cairo. After returning to Cairo, Guy was initially assigned another horse but was eventually reunited with his beloved mare. The Australian Light Horse did more training in the Sinai desert and their Waler horses, which had been trained in the Australian bush, had stamina and strength needed to withstand the 50 degree summer temperatures and the long treks, mostly at night. The 12th Light Horse regiment were part of the battle of Gaza. When this failed, their General Chauvel decided to trek inland and take the city of Beersheba. After riding for several nights and trying to sleep by day, on arrival at Beersheba the soldiers had to fight all day. As the sun was about to set, the 4th and 12th Light Horse regiment were ordered to charge for the town and its precious ancient Wells.

Supported by the 5th Light Horse in reserve, 800 Light Horsemen defeated 4,000 entrenched Turks in an audacious charge over several kilometres and jump the trenches before galloping into town and take the Wells.

Midnight and Lieutenant Guy Haydon, were two of the casualties that day. A bullet went through Midnight's stomach as she jumped the Turkish trench, continued through the saddle and lodged in Guy's back, millimetres from his spine. Only Guy eventually recovered.

Most of these photographs were taken and printed in the field in October and November 1917, and form part of the Haydon family archive. Lieutenant Barney Haydon took the bulk of these, many on Velox paper in primitive printing conditions, and both did pencil notation before archiving them in albums. This is their first publication to help celebrate the last great cavalry charge 100 years ago by the Australian Light Horse.

AN EPIC FROM EGYPT.
By TROOPER BLUEGUM.

It ain't no use a-swearin',
  It ain't no good to fret;
There's little gained by grousin'
  Or getting all upset.
This wilderness is rotten —
  All flies, and dust, and tears,
But the Israelites they stuck it
  For years and years and years.

The Willie-Willies choke yer,
  The dust-storms get yer down;
The red sun robs yer beauty
  And burns yer black and brown.
The drought is something shocking:
  The thirst, our squadron fears,
Can only be abolished
  By beers and beers and beers.

But war won't last for ever,
  This scrap'll soon be done,
An' we'll have done our little bit
  A-strafing o' the Hun.
An' when we get back home again,
  An' meet our little dears,
All thought of Egypt will be drowned
  In cheers and cheers and cheers.

(From "London Opinion".)

SEASONS Greetings

DAVID BARKER

EGYPT-1916

A Christmas message from Trooper Bluegum.

KILLED & WOUNDED?

❊ B Squadron, No. 3 Troop ❊
12th Regiment Australian Light Horse, Australian Imperial Force

EUROPEAN WAR, 1914-15.

W. A. SHEA...
130 PHILLI...
SYDN...

"Our original Troop"
With Guy's notation 1917
Guy Haydon, fourth left back row
Barney Haydon, third in 2nd row

Guy Haydon, 12th Light Horse

Barney Haydon
12th Light Horse

"Stinno"
Corporal Stinson can't find his saddle

Trooper Idriess
With the 5th Light Horse

5th Light Horse
Trooper Ion Idriess, left of troop.

5th Light Horse Machine Gun Troop.

My mate Foot
Corporal Foot, 5th Light Horse

Lieutenant Hampton, Colonel Single, Lieutenant Guy Haydon, and
Colonel McIntosh (left to right, in right-side of photograph)

Third Battalion, on the move

Midnight, when rescued from the 11th Battalion

5th Light Horse with their camel transport

Rounding up captured camels

Taube!
Bombing our lines.

Woe!
Horse killed by Taube at Malah.

Drawing water
This time a Bedouin Well

Our Lines
With defensive trenches

12th in the trenches, Deuidar

Barney's group, having an "intimate" dinner

Australian Rules!
An Australian Rules team of which Barney was playing in

Barney
Taking a blinder!

12th Light Horse Polo Team
Guy Haydon, Jack Davis, Basil Cappe, and Len Williamson

We'll go by horse...

A Desert stream

Preparing for everything

Fagged!
4th Light Horse, after a skirmish.

Shrapnel in the field

Guy Hayden, before the Stunt

TO BEERSHEBA 1917

BREAKING CAMP

The Desert Column

Camel transports
With 12th Light Horse blankets, to Kantara

Indian Cavalry

Indian Artillery unit

To Beersheba
Our mascots, little Egyptian pups.

Infantry
With the 12th Light Horse.

German Taube, crashed in our lines

After the March across the Desert

Turkish prisoners taken by the 6th Battalion

October 1917
Part of the 3rd Troop, B Squadron

At rest
12th Light Horse

44

Meeting some of Lawrence's Bedouins

Captured German fighter plane

Guy with a captured German aviator

Outpost in the Desert

October 1917
Behind their lines

Our Helio group

Posting day in the Desert

At the Scrap!

29 October
6th Battalion

TO BEERSHEBA 1917

THE STUNT BEGINS

12th Light Horse lines.

Our blokes

4th Light Horse
Resting before the Charge

The Stunt begins
The 4th Light Horse about to go

Captured photograph showing Beersheba defences October 30

Letter written by Guy Haydon from Cairo Hospital

You will know from my Cable that I am in Hospital here with a rather nice wound, a bullet about 2 inches to the left of the crupper bone. I will try and give you an account of exactly what happened from the start of operations until I was knocked out.

We left our Camp at Tel-al-fara on the night of the 29th at 5 p.m. and marched to a place called Essani reaching there about 11 p.m. and camped there the night and all next day. About 3 p.m. an enemy Plane came along but was driven off by our planes. At 5 p.m. we moved off again and marched to Khalassa reaching there at 10.30 p.m. and camped. At 2 p.m. the following day the enemy plane again endeavoured to fly over our lines but was attacked by two of our Bristol Fighters and much to our satisfaction they succeeded in shooting her down with their machine guns.

At 5 p.m. we were off again and marched all night and on until 9 a.m. next day when we halted in some broken country 4 ½ miles east of Beersheba. The previous instructions were that the mounted troops were to attack Beersheeba at 10 a.m. and we all quite expected to do so as the infantry were due to attack on the other portions of their line at that time, but 10 a.m. came and went and nothing doing, everyone wondered what could have gone wrong. Had the Infantry failed or had the attack only been postponed for a few hours.

Our Brigade was in reserve and we knew that if any hot job happened along, we would get it. At 4 p.m. orders came to mount and we marched along to within 3 miles of the town until we could go no further without being in full view, then we got the shock of our lives, the order came back "All pack horses, excepting Hotchkiss rifle packs, fall out and remain behind".

Then followed a few moments later the order,

"The I2th. & 4th. L.H. Regiments will charge Beersheeba on Horseback, the town is to be taken at all costs" and five minutes later we were on the way.

We trotted for the first 2 miles then the Turks opened fire on us from a line of redoubts about half a mile out from the town and we could hardly hear anything for the noise of their rifles and machine guns. As soon as their fire started we galloped, and you never heard such awful war yells as our boys let out, they never hesitated or faulted for a moment, it was grand.

Every now and again a rider would roll off or a horse fall shot but the line swept on. As we neared their trenches, our men were falling thicker and thicker and the pace became faster. 30 yards from their trenches were some old rifle pits and as soon as my eye lit on them I wheeled my horse round and yelled to the nearest men to jump off, let their horses go and get into the pits and open fire. Just previously I had seen Major Fetherstonhaugh's Horse go down killed, the Major get up and run for cover only to fall again shot through both legs. A few seconds afterwards a bullet hit me high up in the left buttock, just under the belt, lifting me clear off my horse and dropping me sprawling on a heap of dirt that had been thrown out of a rifle pit, and I rolled down into the pit and into safety.

But all this time, really only a few seconds, the charge went on, men raced their horses through and over the trenches and while some of us were still engaged in hand-to-hand fighting in the trenches, the remainder had charged through the town and went on to the high ground a mile beyond. The town was ours.

It is impossible to describe the charge, I was talking to a British Cavalry Officer in Hospital who had arrived 3 days previously from France, he went to France with the first batch of English Cavalry and

The Charge
It looked like this

had been there ever since, and he said "I have seen every action in which the British Cavalry have taken part, but the charge of the L.H. at Beersheeba yesterday, is the finest thing that I have ever seen mounted troops do." Our Brigadier received a wire from the G.O.C. congratulating him on the brilliant work his Brigade had done.

It is impossible to describe one's feelings, but for myself although it is the heaviest fire I have been under, I never felt less afraid, and I was terribly disappointed in being shot before reaching the town.

We took 2,000 Prisoners and their trenches were full of dead. Two Regiments of the first Brigade also had a charge, but they were further round on our right and we didn't see them, anyway we had the town before they got there. I will give you my experiences from the time I was hit until my arrival here.

I lay in the hole for about 2 hours listening to poor devils groaning all round me, and then an M.O arrived with a lantern and some sandcarts, he planked the lamp down near me and the stretcher bearers brought in the wounded from all points of the compass to be dressed, after being dressed the worst cases were loaded into carts and sent off to the Ambulance, 4 of the poor chaps died there within a yard or two of me, but it did not worry me, I had got past worrying.

At last there was only myself and 1 man left and we had to lie there all night. One of the boys got me a blanket off a dead horse but it was terribly cold, and I shivered all night long and in the morning my wound was so stiff that I couldn't move. About 7 a.m. a sandcart arrived and I was taken to the Field Ambulance where my wound was carefully dressed, then, we went per car to Beersheba then on to the rail head to a big casualty clearing station, where we spent the night. At 9 a.m. we were loaded onto the Hospital Train and reached El-Arish about 2.30 p.m. that afternoon. We spent the night there and left the next day at 12 a.m. for Kantara which we reached about dusk. The next day at about 11 a.m. we boarded the train for Cairo and finally reached the 14th A.G.H. (The best spot on this side of the water). At 4 p.m. today I was X-rayed and the bullet was located about half way up my back and about an inch to the left of my spine, it must have hit a bone and turned at right angles, otherwise it must have gone right through my belly a wonderful streak of luck, am not suffering much pain and don't know when they will operate on me, but hope it will be soon as I don't want to be stuck in here any longer than I can help.

I can't get any correct estimate of the killed and wounded in the 12th yet but may hear in time to put it in this letter yet.

LATER. Was operated on the day before yesterday and bullet removed, am sending you the bullet for a Christmas Present by the same Officer who takes this letter. Am having very little trouble with my wound except at night when it aches a lot, but it is nothing to what some of the poor devils have to suffer. Poor old Nearguard was killed, I was awfully sorry about him, he was such a good Soldier, absolutely fearless. Roy Whiteman and Maclean both have commissions. Roy did splendidly, so well in fact that he was paraded to the Divisional General Hodson and actually promoted on the field of battle. Major Fether-stonehaugh got a D.S.O. He is the bed opposite me. His wife nearly went mad when she heard about it. As far as can gather, there, there must have been about 27 12th L.H. killed in the charge and about 15 wounded. A very high percentage of Killed.

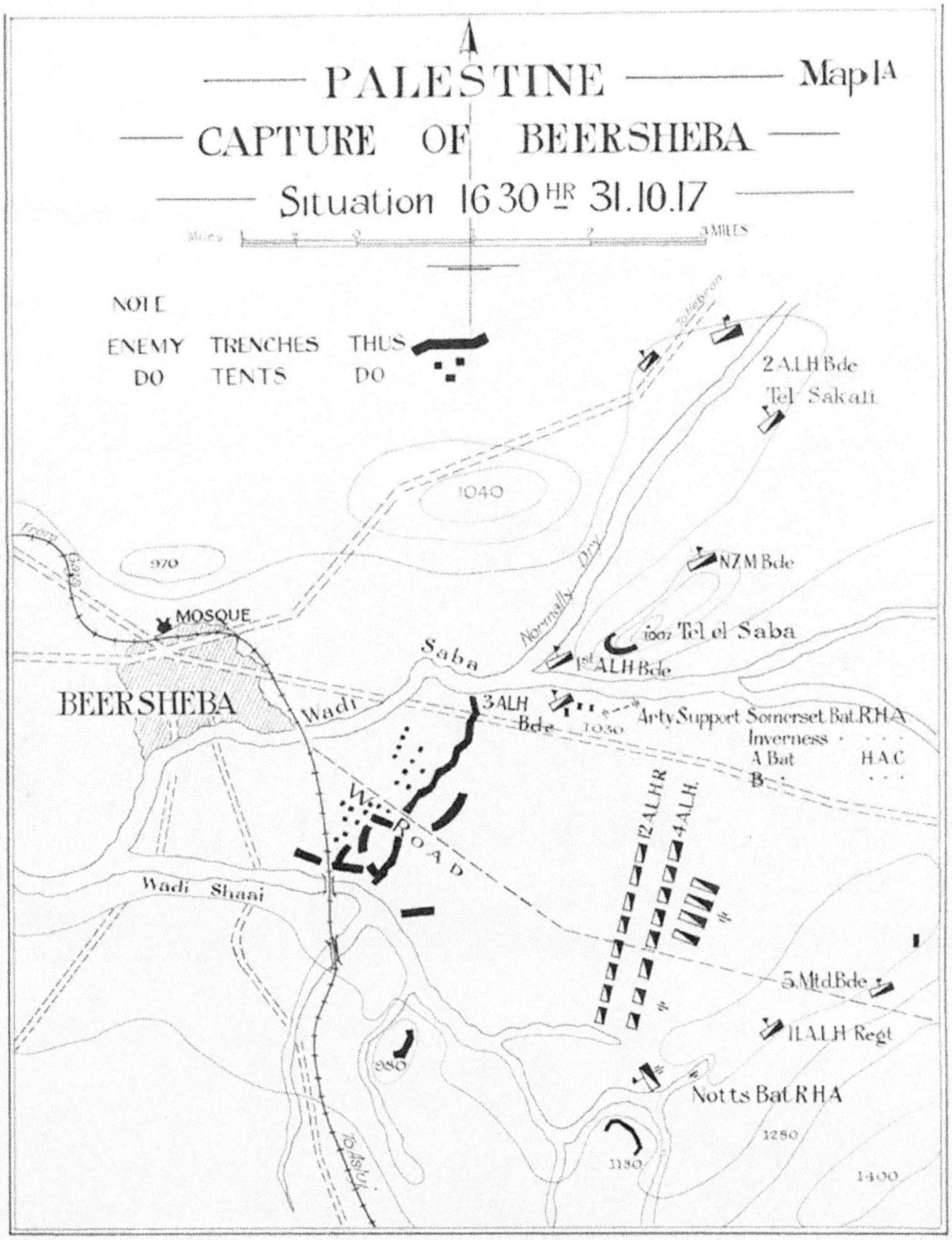

PALESTINE — Map 1A

CAPTURE OF BEERSHEBA

Situation 16 30 HR 31.10.17

Miles 1 0 1 2 3 MILES

NOTE

ENEMY TRENCHES THUS
DO   TENTS   DO

2 A.L.H Bde

Tel Sakati

1040

970

Teerm Sara

MOSQUE

NZM Bde

1007 Tel el Saba

Saba

1ˢᵗ ALH Bde

BEERSHEBA

Wadi

3 ALH
Bde

1030

Arty Support Somerset Bat. R.H.A

Inverness

A Bat   H.A.C

B

Wadi Shaai

W ROAD

12 A.LH R 4 A.L.H.

5. Mtd. Bde

I.L.A.L.H Regt

980

Notts Bat R.H.A

1150

1250

1400

63

First snap of the town

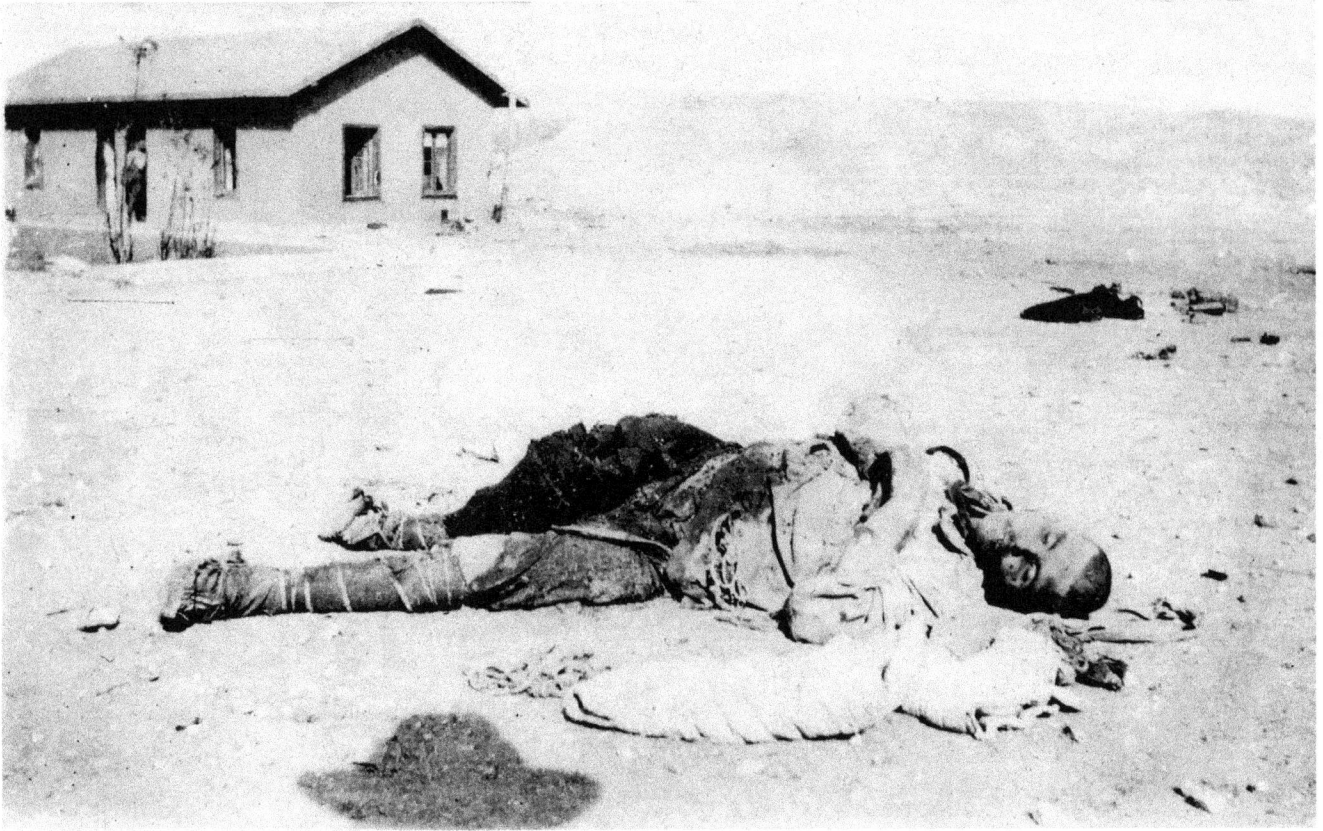

The dead at Beersheba

Dead German Machine Gunners

Turkish Artillery

Unsaddling

4th Light Horse, at rest.

Over 2000 prisoners...

Amongst the ruins, Beersheba

Once a railway station.

Water tanks, blown apart

Atop the Wells of Beersheba.

BEERSHEBA 1917

NOVEMBER 1ST

Collecting our dead

Collecting the saddles from our dead horses

Beersheba
Trenches taken by the 4th Brigade

The Day after the Stunt

Light Horse Helio station, Beersheba

Ambulance transport arrives

Rounding up captured artillery

A hidden Howitzer

Turkish aerodrome, Beersheba

Turkish prisoners.

Prisoners
With the Australian Light Horse.

Some Turkish soldiers we had fought at Gallipoli

Austrian, German and Turkish prisoners

Our Smithy making shoes.

The remains of the battlefield

Bringing in the dead

Burying the dead

Captured German aircraft, ready to go

General Allenby meets our Troop

General Allenby acknowledges the Bengal Lancers

Barney (right) with his mates

November 3
Barney is O.K!

Guy in Cairo hospital, November 1917

Extract from The Desert Column by Ion Idriess

As the regiment mounted, several of us were hurriedly detailed to remain on an observation duty. The brigade galloped off and soon were astride the road. We, on observation, climbed a hill and watched the battle for the remainder of the day. It was all hazily distinct so far as the eye could visualize though obliterated again and again by rolling clouds of dust. Away to the left the New Zealand Mounted Rifles were having a hard fight to take the Tel el Saba redoubts. The machine-gun fire just roared from down there, our artillery all along the line were thundering at the German machine-gun nests. As the afternoon wore on we watched the 1st Light Horse Brigade fighting their way around the flank of a redoubt. Taubes were roaring all over the fortifications, the plain, the wadi and the ridges, their heavy bombs exploding in series of smashing roars. Through the glasses, we watched them bombing Chauvel's and Chaytor's head-quarters four miles away where the generals directed the battle. I wondered what their thoughts were for all the operations, apart from the dust, were spread plain before them. Chauvel must have been terribly anxious as time wore on for if we did not take Beersheba by nightfall then we must retire to water thirty miles away and the infantry divisions now in action right to Gaza would be in a terrible fix. We saw that grim work would soon be doing on Tel el Saba as the 3rd Brigade came galloping up to reinforce the En Zeds. We watched excitedly as we saw the New Zealanders, like little men, advancing in short rushes. Then farther along, the 1st Light Horse Brigade began advancing in bent-backed rushes. Machine-gun, rifle, and artillery fire increased in fury. Then we caught the gleam of bayonets—we strained our eyes as one line of men were almost at a trench, they were into it— faintly we heard shouts as line after line surged on. Quickly the firing from Tel el Saba itself died down. Then we saw it was taken! We just laughed— we were jolly glad. Time rolled on. The outer defences were ours but Beersheba still held out. It was almost sundown, and by Jove we wondered what was going to happen next. The 9th Regiment and its machine-gun squadron were heavily bombed, the New Zealanders got hell from the taubes, while others flying low spread death among the 8th Light Horse. We heard after-wards that their V.C. colonel was among the killed. Then someone shouted, pointing through the sunset towards invisi-ble headquarters. There, at the steady trot, was regiment after regiment, 279 squadron after squadron coming, coming, coming! It was just half-light, they were distinct yet indistinct. The Turkish guns blazed at those hazy horsemen but they came steadily on. At two miles distant they emerged from clouds of dust, squadrons of men and horses taking shape. All the Turkish guns around Beersheba must have been directed at the menace then. Captured Turkish and German officers have told us that even then they never dreamed that mounted troops would be madmen enough to attempt rushing infantry redoubts protected by machine-guns and artillery. At a mile distant their thousand hooves were stuttering thunder, coming at a rate that frightened a man—they were an awe-inspiring sight, galloping through the red haze— knee to knee and horse to horse—the dying sun glinting on bayonet-points. Machine-guns and rifle-fire just roared but the 4th Brigade galloped on. We heard shouts among the thundering hooves, saw balls of flame amongst those hooves—horse after horse crashed, but the massed squadrons thundered on. We laughed in delight when the shells began bursting behind them telling that the gunners could not keep their range, then suddenly the men ceased to fail and we knew instinctively that the Turkish infantry, wild with excitement and fear, had forgotten to lower their rifle-sights and the bullets were flying overhead.

The Turks did the same to us at El Quatia. The last half-mile was a berserk gallop with the squadrons in magnificent line, a heart-throbbing sight as they plunged up the slope, the horses leaping the redoubt trenches— my glasses showed me the Turkish bayonets thrusting up for the bellies of the horses—one regiment flung themselves from the saddle—we heard the mad shouts as the men jumped down into the trenches, a following regiment thundered over another redoubt, and to a triumphant roar of voices and hooves was galloping down the half mile slope right into the town. Then came a whirlwind of movements from all over the field, galloping batteries— dense dust from mounting regiments—a rush as troops poured for the opening in the gathering dark— mad, mad excitement— terrific explosions from down in the town.

Beersheba had fallen.

THE
DESERT COLUMN

ION IDRIESS